Igor S. Makarov

REFORM

OF

MODERN SCIENCE

POLITICS

ECONOMICS

Reform Science Center

Makarov, Igor S., 1935-

Reform of Modern Science. Politics. Economics/Igor S. Makarov
This book is an enlarged compilation of articles published online in 2011

This book has been composed in Times New Roman with the use of the OpenOffice.org 3.0 programs

This book was printed from the camera-ready copy provided by the author

ISBN

All orders and remarks should be sent to the following address:
P.O. Box 461, Haifa 31003, Israel

Content

Preface

This work has been prepared by the author's half-life-long research in physics initiating the reform of modern physics and three recent articles on the subjects of this book published online. As the reform of modern science is a new and actual theme nowadays, it was decided to collect those three articles and publish them in a small book thus taking advantage of their synergy effect and drawing more attention to the program of our Reform Science Center. I am not a professional either in politics or economics and therefore would welcome any remarks and comments from readers concerning the subject matter of the book and other relevant issues.

January 2012, Haifa, Israel

Introduction

As events imply, civilization is experiencing now a general ideological crisis inexorably developing and threatening the very existence of humankind. Underlying this crisis, in our time of science, is mainly the crisis of modern science. Indeed, modern science, despite its stunning technological progress and huge investments in research and development programs, has almost ceased to function as a means of cognizance of nature, which is supposed to be the main function of humankind and the warranty of its existence on earth. Thus, for civilization to survive, it is necessary first to reform modern science.

G. Hegel seems to have been the first to express serious criticism of modern science, while showing the way to overcoming the inconsistency of its methods. He showed that every science is a system of concepts and should be stated in accordance with the logical theory of systems presented in his work on dialectical logic [1]. The main ideas of that work have been interpreted in modern terms largely due to our recent work in physics [2]. That break-through made it possible to develop a general methodology of the reform, stated in the first part of this book, and, taking advantage of our drafts in economics made late in the 80s, start reforming humanities, first of all, politics and economics, the short accounts of which are stated in the second and third parts of this book, respectively.

References

1. G. Hegel. The Logic. Encyclopedia of the Philosophical Sciences, vol.1. Clarendon Press, 1874.
2. Igor S. Makarov. A Theory of Ether, Particles and Atoms. Second Edition, 2010. ISBN-13: 9 781441 478412 (www.amazon.com). Online: http://kvisit.com/S2uuZAQ.

1 - REFORM SCIENCE AND RESEARCH

Abstract

The new method, *'the method of systematic intuition',* based on dialectical logic and successfully used in our research in physics, is generalized and made applicable to the revision of modern science in general, thus giving birth to a new science – *the reform science.*
The reform science has *a structure* common for all branches of science, which allows to introduce *the classification of concepts*, thus purifying, perfecting and organizing the whole science. The reform science consists of three parts, called *Medium, Population* and *Associations,* each with a different logic, that of *transition, reflection* and *evolution,* respectively. The state of the reform science is described by three tables of concepts kept in the Reform Science Bulletin; the research works of the reform science are kept in a dedicated Archive. Every stage of research consists of a paragraph of speculation and a formal statement of the concept. The whole research is a series of such stages, every new stage starting with a speculation about the previous one; the first concept being the origin of the science, its fundamental contradiction suggested by the speculation about the science itself. The research starts with finding the Origin of the branch and proceeds with revealing its Essence, working out its Project and fulfilling its Realization.

Content

Introduction

Now that civilization, enlightened, agitated and inspired by the current technological revolution, expects the proper changes in social and political spheres and, to survive, should be organized as a whole, the role of modern science as an influential source of ideology is of paramount importance. Modern science, however, despite its stunning technological achievements, is experiencing a deep crisis and unable to develop into *the spiritual guide* of society, which it is potentially. Instead, paradoxical as it is, modern science seems to present now *the main threat* to society and should be reformed and organized first.

As to the origin of the above crisis, we should note that the cognizance of the Nature and human society is not the province of science alone. Historically, it has been developing in three spheres: religion (The Unity), philosophy (The General) and science (The Specific). For common success, these three spheres should be in harmony, as was at the time of Aristotle, otherwise there arises a crisis. The present crisis of science originated mainly at the time of Renaissance, when the great success of exact sciences gave birth to the illusion of science being the only true source of knowledge, and its further development disturbed the initial tripartite harmony. Thus, to overcome the crisis of science, it is necessary restore the harmony of the above three spheres.

To reform modern science, it is necessary to have the proper ideology in this respect. Providentially, Hegel's works, critical of scientific methods of his time, convincingly suggest that such an ideology should be some systematic theory based on dialectical logic [1]. No wonder that suggestion prompted Karl Marx to undertake his own interpretation of Hegel's philosophy and apply it for his life-long research in economics [2]. However, despite the importance and great consequence of the latter, it has remained unclear whether it was worthwhile and possible to proceed on that way with other sciences and economics itself. Fortunately, in the course of our research in Systems Theory and Theoretical Physics [3], we have succeeded in our own interpretation of Hegel 's Logic and found solution to the above enigma. Our research initiated the reform of modern physics and paved the way to the reform of modern science in general, which was confirmed by our recent works in politics and economics.

1. Reform science

1.1. Method

In this work, the new method of research, first developed and applied in our research in physics [3], is generalized and made appli-cable to the revision of modern science in general, thus giving birth to a new science – *the reform science*. This method is based on Hegel's dialectical logic and may be called *'the method of systematic intuition'.* Although this method can potentially solve any correctly stated problem, it is not a clear-cut one easy to use in all cases; it cannot be formalized and should be applied with the highest extent of creativity.

According to this method, every stage of research consists of two phases, a paragraph of *speculation* and *a formal statement of the concept*, the former suggesting the latter by necessity, any concept corresponding to *an entity*. The whole research is a series of such stages, where any new statement is analyzed by a further speculation suggesting a new statement and so forth until the end. The first concept is the beginning of the reform science reflecting the origin of the research object; it is *a fundamental contradiction* revealed by the speculation about the nature of the object. Thus the development of the reform science follows the development of the research object. So, unlike modern science where the terms 'science' and 'research' have generally different meaning, in the reform science they mean the same.

1.2. Structure of the reform science

The reform science consists of three parts, that may be called *Medium, Population* and *Associations,* each with a different logic, that of *transition, reflection* and *evolution,* respectively. Unlike modern science that is actually a collection of research works and theories in a particular field, the reform science keeps only the research works recognized as reform science works (*the sources*) and, in addition, the records of *the state* of the reform science in every particular field.

The state of the reform science is described by three tables of concepts, one for each part: Table 1 (Medium) and Table 2 (Population), each containing six rows and four columns called *Thesis, Antithesis, Synthesis and Quality,* and Table 3 (Associations) containing eight rows and nine columns called: 1- *Species,* 2-8 – *Substructures,* 9- *Quality,* as shown in Sec.2. The distinguishing quality of the synthesized entities of Table 1 and Table 2 and of the species of Table 3 are registered in the cells of the last columns.

Like any modern science, the reform science is actually a system of concepts corresponding to various *entities* characteristic of the research object. But, in contrast to modern science, the reform science has *a structure* common for all branches of science, which allows to introduce *the classification of concepts*, thus purifying, perfecting and organizing the whole science. So the reform science is the truly systematic science based on the logically consistent system of concepts. Owing to this property, the reform science is able to sort out the existing concepts, right and generalize them and find the proper meaning to them, and, when necessary, introduce new concepts.

The reform science is a thoroughly *theoretical* science, which corresponds to Hegel's dictum that *'truth cannot be observed, it can only be thought'* [1]. Thus the reform science cannot be developed or verified experimentally; on the other hand, it takes into consideration all achievements and the whole experimental base of modern science and can provide the true explanation to every experimental fact. The reform science realizes the goals advanced by modern science.

2. Research

Before starting the reform of a particular branch of science, the researcher is recommended to get acquainted with the works mentioned above, at least. Then he may start his research in his own field, using those works for *reference by analogy*.

2.1. Part 1. Medium

The structure of Part 1, with its classification of concepts, is presented by Table 1. Medium. In this table, the columns A, B, C are intended for the concepts and their brief description, while the column Q is for the qualitative characteristics of the corresponding *entities* of the column C. Every concept of Table 1 is classified as SC-1ik, where SC (SCIENCE) is the common two-letter abbreviation of the name of a particular science (PH for physics, BI for biology, etc.), i – the column letter (A, B, C), k – the row number (1-6). So the researcher must fill in all the cells of the table with the proper concepts and qualitative characteristics.

The research starts with a paragraph of speculation to suggest an entity introduced by the statement of its concept SC-1A1. This step may prove the most difficult, because this concept has no predecessor and, as mentioned above, should be determined by a speculation about the nature of the research object itself, its original fundamental contradiction to be resolved by the whole research.

Then the research proceeds with a speculation about the entity SC-1A1 to suggest its *transition* to its *dual* entity marked by the concept SC-1B1. After that the research proceeds with a speculation about the two preceding entities, *the thesis* and *the antithesis*, to suggest their *synthesis*, a new entity marked by the concept SC-1C1. The latter has its specific *quality* to be registered in the cell SC-1Q1. The speculation about the entity SC-1C1 generates the entity SC-1A2 to be transited by a new paragraph of speculation to the entity SC-1B2, and so forth until determining the entity SC-1C6 and its quality SC-1Q6. The entity SC-1C6 is *the essence* of the science, its concept being central for the whole theory. Every step is *a discovery* revealed by *intuition* rather than found by a formal work of intellect.

Table 1. Medium

A Thesis	B Antithesis	C Synthesis	Q Quality
SC-1A1 (Origin)	SC-1B1	SC-1C1	SC-1Q1
SC-1A2	SC-1B2	SC-1C2	SC-1Q2
SC-1A3	SC-1B3	SC-1C3	SC-1Q3
SC-1A4	SC-1B4	SC-1C4	SC-1Q4
SC-1A5	SC-1B5	SC-1C5	SC-1Q5
SC-1A6	SC-1B6	SC-1C6	SC-1Q6 (Essence)

2.2. Part 2. Population

Part 2 is similar in many respects to Part 1. Its structure is presented by Table 2 similar to Table 1, and its concepts are classified similarly as SC-2ik. Instead of transition to the opposite, from thesis to antithesis, there takes place here their mutual *reflection* leading to their synthesis, the birth of a new *creature* which settles *the conflict* between its two constituent entities and is characterized by its specific quality. This part of the research starts with a paragraph of speculation about the concept SC-1C6 and ends with the concept SC-2C6 which, as suggested in [3], may be called *the Project.*

Every statement must again be preceded and necessitated by the proper speculation about the statement of the preceding step. As a result, this stage of research generates a series of six creatures, SC-2C1 to SC-2C6, of increasingly higher order and quality, populating the Medium.

Table 2. Population

A Thesis	B Antithesis	C Synthesis	Q Quality
SC-2A1	SC-2B1	SC-2C1	SC-2Q1
SC-2A2	SC-2B2	SC-2C2	SC-2Q2
SC-2A3	SC-2B3	SC-2C3	SC-2Q3
SC-2A4	SC-2B4	SC-2C4	SC-2Q4
SC-2A5	SC-2B5	SC-2C5	SC-2Q5
SC-2A6	SC-2B6	SC-2C6	SC-2Q6 (Project)

2.3. Part 3. Associations

This part is different. Its classification of concepts is presented by Table 3 which illustrates the evolution of the *species* from the simplest one to the most complex. The classification of the species is given in the first column as SC-3-i, where i=0, 1, 2, ...7 is the number of the row. Columns A-G are intended for the description of *substructures,* which are classified as SC-3ik, similar to the classification in the preceding tables. The evolution of species happens by developing new substructures of increasingly higher order. The simplest species, SC-3-0, has no substructures, while the species SC-3-i, i=1, 2, ...7, have increasing numbers of substructures up to the most complex one, SC-3-7, with seven substructures. Similar substructures belonging to different species are generally different and need different classification, as shown in Table 3, but this subtlety may be omitted in the beginning. The last column is intended for the qualitative characteristics of different species.

The research of this part starts with a paragraph of speculation about the concept SC-2C5 and SC-2C6 to suggest their *merger* into the species SC-3-0, *the fundamental component of substructures.* A speculation about the latter should suggest the species SC-3-1 consisting of the substructure SC-3A1. Further speculation should expose the internal contradiction of SC-3-1 necessitating its development, *evolution,* by generating a new substructure, SC-3B2, which adds to SC-3B1 to make the species SC-3-2. The research proceeds further until generating the substructure SC-3G7 which adds to the preceding six substructures, SC-3A7, SC-3B7, SC-3C7, SC-3D7, SC-3E7, SC-3F7, to make the most perfect species SC-3-7, *the Realization.* To complete in rough the research, it is necessary to show the concept SC-3-7 to be indeed the realization of the concept SC-2C6.

Table3. Associations

Substr. → Species ↓	A	B	C	D	E	F	G	Q Quality
SC-3-1	SC-3A1							SC-3Q1
SC-3-2	SC-3A2	SC-3B2						SC-3Q2
SC-3-3	SC-3A3	SC-3B3	SC-3C3					SC-3Q3
SC-3-4	SC-3A4	SC-3B4	SC-3C4	SC-3D4				SC-3Q4
SC-3-5	SC-3A5	SC-3B5	SC-3C5	SC-3D5	SC-3E5			SC-3Q5
SC-3-6	SC-3A6	SC-3B6	SC-3C6	SC-3D6	SC-3E6	SC-3F6		SC-3Q6
SC-3-7	SC-3A7	SC-3B7	SC-3C7	SC-3D7	SC-3E7	SC-3F7	SC-3G7	SC-3Q7 Realization

3. The research work formats

As mentioned above, the reform science publishes both the state of science and the sources, the research works themselves. To this end, there should be two kinds of media: *The Bulletin of the Reform Science,* possibly a quarterly journal, publishing the state of different branches of the reform science in the form of above three tables of concepts, and *The Journal of the Reform Science Archive,* publishing the research works recognized as the likely sources of the reform science to be kept in a specialized library, *The Reform Science Archive.*

The structure of the reform science and its logic outlined above require a definite format of the research work to be published. The first requirement is that the research work should include, at least, one of the whole Part 1 or Part 2 or, possibly, Part 3 of a particular branch of science, because otherwise it would be difficult to use the self-correction property of the reform science and estimate the correctness of the whole work. As to Part 3, which may prove to be the most difficult one, it is admissible to proceed with the research in this part and its publication by substructures, in accordance with the logic of the part. In this connection it is recommended to use the definite research work formats stated below.

3.1. Format of Part 1

Title of the work (Name of the branch. Part 1)
Introduction: a short overview of the branch showing its contradictions and the necessity of the reform.
Part 1. Title of Part 1 (Name of the *Medium* – as suggested by the concept SC-1C3)
1. Name of the branch
    A. Speculation about the origin of the branch suggesting the concept SC-1A1.
    B. Speculation about SC-1A1 suggesting the concept SC-1B1.
    C. Speculation about SC-1A1 and SC-1B1 suggesting their synthesis SC-1C1 and quality SC-1Q1.
2. Name of the concept SC-1C1
    A. Speculation about SC-1C1 suggesting SC-1A2.
    B. Speculation about SC-1A2 suggesting SC-1B2.
    C. Speculation about SC-1A2 and SC-1B2 suggesting their synthesis SC-1C2 and quality SC-1Q2.
3. Name of the concept SC-1C2 (A, B, C)
4. Name of the concept SC-1C3 (A, B, C)
5. Name of the concept SC-1C4 (A, B, C)
6. Name of the concept SC-1C5
    A. Speculation about SC-1C5 suggesting SC-1A6.
    B. Speculation about SC-1A6 suggesting SC-1B6.
    C. Speculation about SC-1A6 and SC-1B6 suggesting their synthesis SC-1C6 and quality SC-1Q6.
Table of concepts (Table 1)

Conclusion
References

3.2. Format of Part 2

Title of the work (Name of the branch. Part 2)
 Part 2. Title of Part 2 (The general name of the *Population* suggested by the concepts SC-2C)
Introduction: a short overview of the Part 1 showing the necessity to proceed with the research
1. Name of the concept SC-1C6
   A. Speculation about SC-1C6 suggesting the concept SC-2A1.
   B. Speculation about SC-2A1 suggesting the concept SC-2B1.
   C. Speculation about SC-2A1 and SC-2B1 suggesting their synthesis SC-2C1 and quality SC-2Q1.
2. Name of the concept SC-2C1
   A. Speculation about SC-2C1 suggesting SC-2A2.
   B. Speculation about SC-2A2 suggesting SC-2B2.
   C. Speculation about SC-2A2 and SC-2B2 suggesting their synthesis SC-2C2 and quality SC-2Q2.
3. Name of the concept SC-2C2 (A, B, C)
4. Name of the concept SC-2C3 (A, B, C)
5. Name of the concept SC-2C4 (A, B, C)
6. Name of the concept SC-2C5
   A. Speculation about SC-2C5 suggesting SC-2A6.
   B. Speculation about SC-2A6 suggesting SC-2B6.
   C. Speculation about SC-2A6 and SC-2B6 suggesting their synthesis SC-2C6 and quality SC-2Q6.
Table of concepts (Table 2)
Conclusion
References

3.3. Format of Part 3

Title of the work (Name of the branch. Part 3)
Part 3. Title of Part 3 (The general name of the *Associations* sug-

gested by the concepts of the first column)

Introduction: a short overview of the Part 2 showing the necessity to proceed with the research

1. Name of the species SC-2C6

Speculation about SC-2C6 suggesting its merger with SC-2C5 giving birth to species SC-3-0 (the fundamental component of substructures) with quality SC-3Q0.

2. Name of the species SC-3-0

Speculation about SC-3-0 suggesting SC-3-1 (substructure-center SC-3A1) with quality SC-3Q1.

3. Evolution of the species SC-3-1

3.1 Name of the species SC-3-1

Speculation about SC-3-1 suggesting SC-3-2 (new substructure SC-3B2) with quality SC-3Q2.

3.2 Name of the species SC-3-2

Speculation about SC-3-2 suggesting SC-3-3 (new substructure SC-3C3) with quality SC-3Q3.

3.3 Name of the species SC-3-3

Speculation about SC-3-3 suggesting SC-3-4 (new substructure SC-3D4) with quality SC-3Q4.

3.4 Name of the species SC-3-4

Speculation about SC-3-4 suggesting SC-3-5 (new substructure SC-3E5) with quality SC-3Q5.

3.5 Name of the species SC-3-5

Speculation about SC-3-5 suggesting SC-3-6 (new substructure SC-3F6) with quality SC-3Q6.

3.6 Name of the species SC-3-6

Speculation about SC-3-6 suggesting SC-3-7 (new substructure SC-3G7) with quality SC-3Q7.

3.7 Name of the species SC-3-7

Speculation about SC-3-7 showing it to be the realization of the Project (SC-2C6) and the solution of the original fundamental contradiction (SC-1A1).

Table of concepts (Table 3)

Conclusion

References

4. Reform Science Center

The reform of modern science can be achieved only by coordinated efforts of the whole scientific community. Thus it would be necessary to set up an *International Center of the Reform Science* to coordinate the research in the reform science the world over. The Center is to become the engine and the embodiment of the reform of modern science. Governed by the Reform Science Council, it would instruct, guide and coordinate scientific activity the world over, while collecting, processing, unifying and publishing its essential results. To function properly, the Center should have a bulletin, a journal and an archive, as mentioned above. At present we have its virtual substitute, The Reform Science Center [4], that can serve as a prototype of the above main project. The Reform Science Center publishes an online Bulletin which covers the *state* of the reform science and *references* of research sources for every branch of science kept in its online Archive.

The Bulletin abides by the following policy:

1. When a research is submitted, the Editorial Board makes its rough estimation and, if recognized as a possible version of the reform science, treats it as a reform science source to be kept in the Archive and available free.

2. If there are two or more different research works in the same field of science, the Bulletin keeps them all as likely reform science sources and strives to resolve the confusion, finding the *right* source.

3. If the confusion is resolved, the Bulletin refers the state of science only to the right source.

4. Otherwise, the Bulletin publishes all the alternatives of the state of science with the respective references.

5. The Bulletin registers the state of every branch of the reform science by publishing Table 1 and Table 2 of concepts with the respective references and Table 3 with references for every substructure separately.

Conclusion

The above text suggests that the whole research of the reform science, in its every branch, is a great endeavor concerned with finding the Origin of the branch, revealing its Essence, working out its Project and fulfilling its Realization. By elucidating the structure of the reform science and the meaning of its cornerstones, this article makes the formerly unimaginable task of reforming modern science much less daunting, much more feasible and perhaps even more fascinating.

References

1. G. Hegel. Encyclopedia of Philosophical Sciences: vol.1 - The Science of Logic; vol.2 - Philosophy of Nature; vol.3 - Philosophy of Spirit. Clarendon Press, Britain, 1874.
2. K. Marx. Capital, vol.1: A Critique of Political Economy.. Amazon.com.
3. Igor S. Makarov. *A Theory of Ether, Particles and Atoms. Second Edition,* 2010. Open University Press. Manchester. UK.
Order: www.amazon.com, ISBN-13: 9 781441478412.
Online text: http://kvisit.com/S2uuZAQ;
ts cover: http://kvisit.com/S2-uZAQ.
4. The Reform Science Center: www.reformscience.org.

2 - SCIENCE OF POLITICS

Abstract

This part of the book develops the science of politics as a systematic body of knowledge. The work consists of three chapters: 1- Human society, 2-The government, 3- The self-government. Chapter 1 starts with the investigation of human being, family and the development of society into the world community characterized by its ideal model, the World-Consistent Nation (WCN), governed by the International Law and expected to develop universal religion, philosophy and science. Chapter 2 investigates the logical connection and specific features of different types of government, from monarchy to republic, the latter proving to be ideally the best government, the Project of the above ideal WCN. Chapter 3 shows the transformation of the republican government into the self-government, first as its unstable form, demo-republic, and then its real stable form – empire. There appear two empires with different ideological orientations, social-private and private-social, dividing the whole world into two spheres of influence, competing with each other and solving best all global problems. The evolution of the empire is actually that of its three institutions: the Assembly - an institution responsible for domestic affairs, the Senate – an institution responsible for foreign affairs, and the Church or another religious institution responsible for ideology and justice; each of them having a tripartite structure to represent the other two. Other nations, when developed to the status of demo-republic, join one of the empire with different extents of affinity creating different associations called solidarity, support, neutrality, culture, commerce and global unions, thus promoting the development of the empire from its initial form Empire-1, to its most perfect form, Empire-7, the Realization of the WCN-Project.

Content

Introduction

The science of politics studies human society as an integral developing entity. In contrast to numerous works of the so-called *political
science* based mainly on the analysis of historical facts, this study is
based on arranging in a systematic, logical order the knowledge
about the main stages of human society, predicting possibly its
future development. The study is stated in the format of reform science described in the first part of the book and therefore starts with
the very origin of society, its element, *the human being.*

Chapter 1. Human society

1.1. Human being

A. As the Bible testifies, the first human being was *a man,* Adam.
Adam is begotten by nature, and therefore is different from it, a spiritual animal, *a spirit-body.* That his duality is one-sided, a dishar-
mony suggesting the existence of another dual type of human being,
a body-spirit, a woman.

B. As the Bible testifies, the first woman was *Eve.* Separately, Adam
and Eve are abstract, but they are begotten for each other, feel attraction and meet, thus creating a natural unity, *a family.*

C. The family is the union of two dually equal partners, in which
each side finds its dual self in the other side, becomes self-affirmed
by that unity, *self-conscious* and *real.* The family is the initial human
reality, *the embodiment of spirit,* the foundation of *life.*

Comments:
The above suggests that it is families, not separate human beings,
that are the real constituents of society and subjects of politics.

1.2. Life of the family

A. Partners of the family are engaged in *private intercourse, a corporal, material* embodiment of life, the life proper; the family struggles for existence and acquires *home;* there appear *children.*

B. The children begotten and raised by the family leave the family that eventually disintegrates. The children make their own families which give rise to new children and new families, and a family to occupy the home of the original family, thus *reviving* the original family as *a home-family.* Via that *social intercourse,* the home-family acquires *a social, moral authority* - a *spiritual* embodiment of life.

C. The private and social intercourse correspond to each other, complement and presuppose each other, and together characterize *the social status* of the family, the *quality* of life.

1.3. Social status

A. The social status suggests *a social environment,* the existence of an indefinite multitude of home-families. As regards their private life, families are hostile to each other, repel each other and separate from each other, which manifests itself in their separate *dwellings.* Different dwellings are occupied by different families which as social and moral families are similar to each other.

B. As carriers of moral authority, families are friendly and attracted to each other. That friendship manifests itself in *common faith and traditions.*

C. The separation of families from each other suggests their attraction, because otherwise there would be no necessity for separation; in a similar way, the attraction of families to each other suggests their separation. Thus the separation and attraction of families are complementing features suggesting the existence of their unity, *a world community,* in which every separate family interacts and communicates with its social environment. The world community is characterized by its *communicability* which depends on both above trends.

Through communication, the world community perceives its *unity* and comes to the question of the purpose of its existence; as the answer to this question cannot be found in the sphere of lay and private issues, it should be sought for in the sphere of spiritual, *religious* and *social* issues; thus the world unity gives birth to religion; as the social sphere cannot exist without the private sphere, the answer to the above question proves to be this: the purpose of humankind's existence is both social (religious) and private, *social-private.*

Comments:

The above-mentioned term "private" relates to the human being within his/her human environment, while the term "social" relates to the whole society perceived as a single entity. Thus the private view of society is an internal, subjective one, while the social view is an external, objective one.

1.4. World community

A. The world community is an indefinite multitude of families. That multitude is both *discrete* and *connected*: it is discrete as it consists of isolated families distinguished by their *family names;* it is connected as any single family is *a social, moral entity* similar in this respect to any other family communicating with its social environment.

B. As to the single family, it is also, like the whole community, both discrete and connected: it is discrete as one family distinguished by its *name*; it is connected as a spiritual, moral family similar in this respect to any other family of *ethnicity,* an *ethnic* family.

C. The confrontation between the world community and the single family is settled in *the ethnic community;* the latter is a multitude of families, like the world community, and, like the single family, it has a family name, that of its *chief,* a male or female representative of the family with the most respectable social status. The ethnic community generates its *ethnic culture and religion.*

1.5. Ethnic community

A. The ethnic community unites families having *ethnic affinity* with its chief and is characterized by its *population.* The families that do not meet the criterion of affinity belong to other ethnic communities separated from the first by *a border.*

B. As the criterion of affinity is disputable and unable to determine the border exactly, some families find themselves belonging to two or more ethnic communities simultaneously, which results in conflicts between ethnic communities. But different ethnic communities are located in places with different *geography* and natural, *territorial* borders, which helps to settle the conflicts [1].

C. The ethnic community separated by territorial borders is *a territorial community, a nation;* it has a characteristic *territorial ethnicity* of its families and a definite border with its neighbors. The nation generates its *national culture and religion.*

1.6. Nation

A. The nation establishes its national laws and regulations, which promotes *national restrictions.* But the nation borders others nations with different laws and regulations.

B. The borders between nations become the cause of uncertainty and hostility. The nation struggles to settle its relations with its neighbors; the latter struggle to settle relations with their own neighbors and so forth; as a result, the national laws and regulations experience *a world mediation* and return to the original nation in the form of a mediated, *universal, international law.*

C. With the formation of the international law, the nation becomes consistent with the world and turns into *a world-consistent nation* (WCN) governed by the international law; the WCN being the *essence* of the whole theory. The international law cannot be expressed verbally, it only *manifests* itself, in *constitutions* in particular. The WCN is a society with an *ideal* unity of three entities: the family, the nation and the international community; it is *a religious community,* where all the lay duties have ideally religious motivations. Accordingly, the WCN generates three spiritual cornerstones of its ideology: science, philosophy and the world religion.

1.7. Summary

Table 1. Human society

A Thesis	B Antithesis	C Synthesis	Q Quality
Man. Spirit-body	Woman. Body-spirit	Family. Embodiment of spirit	Life. Self-consciousness
Private intercourse. Corporal side of life	Social intercourse. Spiritual side of life	Social status. Quality of life	Self-affirmation of life
Separation of families. Family names	Attraction of families. Faith, traditions	World community. Communicability	Birth of religion
Multitude of families	Single family. Ethnicity	Ethnic community	Ethnic culture and religion
Chief of community. Ethnicity	Border of ethnicity. Geography	Nation. Geographical borders	National culture, religion
National laws and restrictions	Interaction across borders. Mediation	World-consistent nation (WCN). International law	Universal science, philosophy, religion,

1. 8. Discussion

The above reasoning may put some questions as follows:

(1) Is the world-consistent nation a real one?
No, it is not; it is an abstract, ideal model of the nation which is supposed to comply ideally with the international law. Ideally, the life of any real nation should be as close to the above model as possible.

(2) If so, what are then the fundamental principles of the international law and the ideal constitution?
As suggested above, the international law cannot be expressed verbally, nor can be the ideal constitution. However the most authoritative documents, such as the Ten Commandments of the Old Testament, the New Testament, the Constitution of Medina [2], do state some fundamental principles of the international law and therefore the ideal constitution.

(3) What is the importance of the universal religion, philosophy and science?
As human knowledge suggests, the cognizance of the world and human society is not only the highest inspiration for humankind, but its main function and the warranty of its existence on earth. That cognizance has been developing in three spheres: religion (The Unity), philosophy (The General) and science (The Specific). For common success, these three spheres should be in harmony, as was at the time of Aristotle, otherwise there arises a crisis, which seems to be the case at present. The present crisis originated mainly in the 18th century, when the great success of exact sciences gave birth to the illusion that science was the only true source of knowledge, and its further development disturbed the initial tripartite harmony. Thus, for civilization to develop peacefully, it is necessary to restore the harmony of the above three spheres.

Chapter 2. Governance

In Chapter 1, we have reached the stage where human society is developed into the world-consistent nation governed by the international law. But that society and its law are still latent, abstractions which cannot be directly expressed in familiar terms. So we should proceed with the study of human society to discover its government.

2.1. World-consistent nation

A. In the world presented by the world-consistent nation, every nation is a *particular* religious community living in particular geographical conditions, having a particular national *constitution* and united by its *devotion* to god.

B. Devotion to god, *a universal deity,* suggests necessity for devotion to *a national deity.* Indeed, there arises a religious family with the highest social status that becomes the leader of the community, *a ruler,* the national deity.

C. The religious community governed by the ruler is *a monarchy,* a type of government based on devotion of the people to the *monarch* considered a *minister* of god. The monarchy, an ideal unity of the nation, enlightens the people through religious dogmas to maintain the unity of the nation.

2.2. Monarchy

A. To rule the nation, the monarch-family surrounds itself with relatives and devotees, creating a circle of favorite families and assigning to them various *administrative functions;* the monarch-family thus sustains itself by the devotion of that *proxy-circle* of families; the proxy-circle functions as *a civil security guard* of the monarch-family, suggesting necessity for *a special security guard.*

B. The monarch-family hands down its power to its closest relatives, establishing *a hereditary succession* of power. The hereditary succession of monarchs gradually degrades the social status of the monarch-family and the devotion of its proxy-circle of families. To protect its sovereignty, the monarch-family does set up the above-suggested special security guard and keeps enforcing its rule on the nation by *coercion.*

C. The monarchy governed by coercion leads to *a tyranny, the tyrant* being the embodiment of the *power* of the nation. The tyranny enforces a habit to *social discipline* necessary to maintain the unity of the nation.

2.3. Tyranny

A. The tyrant-family rules for the sake of its own power. By thus cynically using its power and pushing its security measures to the extreme, it ceases to be a religious family and deprives itself of the devotion of the people and that of the proxy-circle of families as well, who start serving simply as *governmental functionaries.*

B. With a degraded status of the ruler-family, there appears a group of respectable and religious persons with a high social status, *aristocrats,* who become national tribunes, exponents of the lofty ideals of the people and leaders of the popular *discontent.*

C. In condition of total discontent, the aristocrats lead the people to overthrow the tyrant family and establish a *collective* form of government, *aristocracy,* "rule of the best", the embodiment of the *morality* of the nation.

2.4. Aristocracy

A. The aristocracy reunites the secular duties of the government with the religious ideals of the people, thus re-establishing the loyalty of the people. The aristocracy is government *at the discretion of the elite,* a group of *wealthy* people.

B. To maintain its power, the governing elite surrounds itself with a circle of the like, wealthy people and relatives, trying to keep the power within that circle; in doing so, without popular control, the governing elite gradually loses its lofty ideals and degrades into a group of *mediocre* people, *wealthy functionaries, oligarchs.*

C. The aristocracy thus turns into *an oligarchy,* "rule of few", the embodiment of the idea of *collective* power. The oligarchs *organize* the government in a way conducive to keep the power within their own circle.

2.5. Oligarchy

A. The oligarchy is the rule by *an organized elite* who, having no lofty ideals for government and no control from ordinary people, rule for their own sake, *an organized collective tyranny.* By separating religion from popular life and substituting it by a set of formal rituals, they turn religion into *an organized religion,* thus *exempting* people from the necessity for sincere consideration of the spiritual content of their daily duties; with that exemption, the life of the nation comes eventually to a conflict with the international law, which results in *a national crisis.*

B. In conditions of crisis, the people find themselves exempt from their loyalty to the government and advance *an organized collective of enlightened people,* who set up *an organized popular movement* against the government; the organized elite now confronts *the organized people.*

C. The struggle between the popular movement and the ruling oligarchy leads to *a revolution* overthrowing the ruling elite and establishing *democracy,* "rule of the people".

History: The classical reference point of early democracy is Athenian democracy established in 507 BC. Originally, it has two distinguishing features: (1) allotment (selection by lot) of ordinary citizens to the few government offices and the courts, and (2) the assembly of all citizens [3].

2.6. Democracy

A. Democracy is a type of government in which all citizens have equal rights to vote and be elected and, making use of this *popular sovereignty*, elect three collective bodies, an administrative council, a legislative assembly and a court, which *collectively, by majority vote,* decide all political issues. These bodies create the precedent of a government with *a primitive hierarchical organization* suggesting the similar organization of the people.

B. Pushing the popular sovereignty to the extreme, becoming "a collective tyrant", "government of "the mob", the democracy loses high religious ideals and the very goal of government, which leads to its decay. The decay of democracy gives rise to *a council of enlightened people* who become carriers of spiritual, religious and scientific ideals of society, critical of democracy. So the administrative council now faces the council of enlightened people – a hierarchical organization of the people.

C. Criticism of the democracy government leads to its disintegration and transformation into *a republic,* a form of collective hierarchical government with *personal responsibility of officials.* Typically for a republic, common citizens elect *a senate* and *an assembly*; the senate consisting of so-called noble citizens (aristocrats) who elect and control administration for offices endowed with supreme power; the assembly consisting of regional representatives who elect administration for civil offices and public affairs. By thus subordinating the authority of common people and aristocrats, the republic reunites the religious ideals and lay duties of the people and ideally presents *the embodiment of the world-consistent nation.*

2.7. Summary

Table 2 below summarizes the above reasoning.

Table 2. Governance

A Thesis	B Antithesis	C Synthesis	Q Quality
Devotion to god, universal deity	Devotion to ruler, national deity	Monarchy - devotion-based government	Religious elevation and unity
Proxy-circle of families. Civil security	Security guard. Special security	Tyranny. Governance based on coercion	Unity enforced by discipline
Government by common functionaries	Aristocrats – enlightened and trustworthy	Aristocracy. Governance based on trust	Education and enlightenment
Circle of enlightened wealthy people. Heritage of property	Circle of wealthy functionaries. Heritage of power	Oligarchy. Governance by self-sustainable circle of functionaries	Power of collective organization
Organized government. Personal authority	Organized people. Collective authority	Democracy. Governance by organized people	Manifestation of people's sovereignty
Hierarchy of offices. Collective responsibility	Hierarchy of organizations. Personal responsibility	Republic. Collective government with personal responsibility	Ideal embodiment of WCN. Ideal model of government

2.8. Discussion

The review of Chapter 2 shows a steady progress of society from monarchy to republic. In that development, every succeeding form of government does not obliterates the preceding one but includes it as its own main principle and mechanism. Indeed, the tyranny does not obliterates the monarchy but includes it as the power of the state;

similarly, the aristocracy includes the tyranny substituting subjugation by fear with subjugation by trust; oligarchy includes aristocracy substituting its collectively organized trust by collectively organized discipline; democracy includes oligarchy substituting its power of organized collective by the power of the organized people; finally, republic includes democracy as the principle and mechanism of the people's representation in the government.

As a result the republic contains all the preceding forms of government – monarchy symbolized and presented by the head of the state, tyranny in the form of laws enforcing discipline on the people, aristocracy presented by the senate, oligarchy presented by the heads of the governmental offices, and democracy presented by the assembly and the representative principle of its election. That conclusion confirms Aristotle's view of the best government which, in his opinion, should be of a mixed type.

Chapter 3. Self-governance

3.1. Republic

Republic seems to be an ideal form of government, but its structure has not been shown explicitly, and it is not clear how to realize it; so republic seems to be rather *a project* of the best government than its real model. To realize that project, we have no other means but to proceed with the analysis of the results obtained.

As we have seen, democracy is a popular government lacking in lofty ideas, while republic is a government controlled mainly by elected aristocracy. In the republic, its senate has a dominant position because it elects the heads of higher offices. However, as the senate is elected by the assembly, the solution of the first should ideally be *in the interest* of the second, the assembly, that is democracy; that means the transition: *republic-democracy*.

As to the democracy, it elects the council which, to perform the proper governance, needs lofty ideals and education, which is the prerogative of aristocracy elected to the republican senate; that means the transition: *democracy-republic*. Thus democracy and republic are dual forms of government that reflect one another, may under the proper conditions turn one into another, and need a mutual interaction with each other to become the true government. Therefore, under proper conditions, democracy and republic start interacting, merging and transforming one into another, which results in the birth of a higher form of government – *a demo-republic,* a *merger* of democracy and republic, in which the assembly elects the senate *and* control its decisions. The demo-republic is the government of competent heads of offices elected and *controlled* by the people, *a self-government.*

3.2. Demo-republic

In the demo-republic, neither the assembly nor the senate has priority over one another, which seems to be the best case. However, if the senate passes laws that do not satisfy the assembly, there may arise a predicament requiring priority of one side. That means that the demo-republic cannot be a stable form of government: it would *alternate* between democracy and republic; that suggests the necessity for *a double demo-republic,* one half having priority in democracy, another in republic - *a bipartisan government.* The demo-republic with a bipartisan government turns into *a double-republic,* a stable self-governance, *an empire.*

3.3. Empire

The empire is a *real* implementation of the world-consistent nation; there emerge two kinds of empire with dual ideologies: some giving preference to private interests rather than to social (religious) ones, *PS-empires,* others to social (religious) interests rather than to private ones, SP-empires.

3.4. Evolution of the empire

The empire evolves by unions with other nations that have reached the level of demo-republic. As demo-republics dispersed over the world have different ideological preferences, unions are concluded accordingly. There are six possible types of union as follows:

(a) *solidarity unions* unite those demo-republics that have a *firm ideological affinity* with the empires of their ideological orientation; these unions are open to all kinds of activity and cooperation;

(b) *preference unions* unite those demo-republics that *prefer* to take side with the empires of their ideological orientation; these unions are open to activities aimed at consolidating the unions;

(c) *neutrality unions* unite those demo-republics that *have no ideological preference*; these unions are engaged in all activities of mutual interest;

(d) *cultural unions* unite those demo-republics that *prefer to take side with the counter-empires*; these unions are engaged mainly in cultural exchange and all activities of less affinity;

(e) *commercial unions* unite those demo-republics that have strong *affinity with counter-empires*; these unions are engaged mainly in commercial activities;

(f) *the global union* unites two global counter-empires, involving them in mutual interaction and activities of global importance. All unions of higher rank can participate in the activities of lower rank unions.

The evolution of the empire is actually the evolution of its three institutions: the Assembly (A) – an institution responsible for domestic affairs (the Specific), the Senate (S) – an institution responsible for foreign affairs (the General), and the Church (CH) or any other religious institution responsible for the ideology (the Unity). Every two of these institutions suggest the third. Indeed, (A, S)→CH because, to decide on domestic and foreign affairs and ensure harmony and unity, the assembly and the senate need the proper ideology provided by the Church; (A, CH)→S because, to decide on domestic affairs and ensure their unity and consistency with foreign relations, the Assembly and the Church need a general approach provided by the Senate; (S, CH)→A because, to decide on general issues and ideology, the Senate and the Church need to consider all the specific issues of social life and therefore should consult the Assembly.

3.5. The global empires

There are two dual global empires, the SP-empire and the PS-empire, with the social-private and private-social ideological orientations, respectively. The global empire, irrespective of its ideological orientation, is the highest stage of the empire presenting the most perfect implementation of the WCN project initiated implicitly by the republican form of governance. The two global empires compete for the spheres of influence, solving in this way all global problems, merging the social and private aspects of life and exposing the best possible interpretation of the international law,

3.6. Global governance

The above theory naturally solves the problem of so-called global governance. Indeed, the above theory shows there is no necessity for global governance at all, because civilization as a whole is a self-governing entity and would develop naturally and best unless impeded. As is suggested above, at higher stages of development, society resorts to self-governance: there appear two empires, PS-empire and SP-empire, that gradually develop by contracting unions with demo-republics as well as between themselves. The ideal scheme of the self-governing world is symbolically shown in Fig.1. Each of the empires has a circle of close allies (solidarity unions) and a circle of less close allies (preference unions) of the respective ideological orientation. There is a group of neutral demo-republics contracting agreements of mutual interest with the empires. The two empires compete with each other for the spheres of influence thus solving all global problems. If one of the empires starts dominating, solving global problems to its own advantage, some of its preference allies change their choice in favor of the counter-empire, thus restoring the global balance of power and justice. This mechanism of global self-governance seems to be quite flexible, able to allow for any contingencies.

3.7. Summary

Table 3. Self-governance

Affinity: Empires	A	B	C	D	E	F	G	Q Quality
Empire-1 (PL-3-1)	PL-3A1							Empire center (PL-3Q1)
Empire-2 (PL-3-2)	PL-3A2	PL-3B2						Solidarity (PL-3Q2)
Empire-3 (PL-3-3)	PL-3A3	PL-3B3	PL-3C3					Preference (PL-3Q3)
Empire-4 (PL-3-4)	PL-3A4	PL-3B4	PL-3C4	PL-3D4				Neutrality (PL-3Q4)
Empire-5 (PL-3-5)	PL-3A5	PL-3B5	PL-3C5	PL-3D5	PL-3E5			Culture (PL-3Q5)
Empire-6 (PL-3-6)	PL-3A6	PL-3B6	PL-3C6	PL-3D6	PL-3E6	PL-3F6		Commerce (PL-3Q6)
Empire-7 (PL-3-7)	PL-3A7	PL-3B7	PL-3C7	PL-3D7	PL-3E7	PL-3F7	PL-3G7	WCN project realization (PL-3Q7)

Table 3 shows the development of the empire from its initial stage, Empire-1, to its highest stage, Empire-7; the first column of the table contains all the stages of the empire; the next seven columns, A-G, correspond to different zones of affinity and contain the lists of demo-republics joining the empire at its different stages for the different reasons as follows: A- the core of the empire; B- solidarity; C- preference; D- neutrality; E- culture; F- commerce; G- global interaction with the counter-empire; the last column (Q) characterizes the distinguishing features of every stage of the empire.

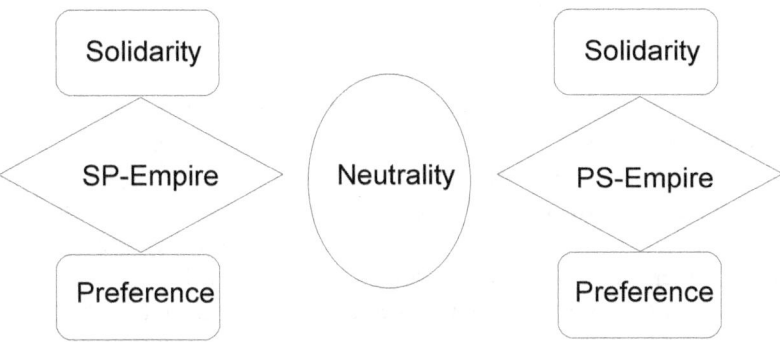

Fig. 1. Global governance

3.8. Discussion

There has been introduced a new concept – demo-republic; is it a
real entity? Yes, it is; it is an initial, simple case of self-governance;
as a separate entity, it is unstable and becomes stable only when
associated with an empire. For example, the Paris Commune, that
emerged during the French Revolution and governed Paris for over
two months (March 18 – May 28, 1871), was clearly a demo-republic even if for a short time. Nowadays, the American states sharing
their power with the federal government is an example of present
stable and flourishing demo-republics.

The existence of dual ideologies, as mentioned above, may result in
different interpretation of some common concepts, such as *democracy*, for example. Indeed, democracy is usually interpreted as equal
rights to vote, as *individual freedom*, the freedom to express personal
views. That seems to be the Western concept of democracy. But
democracy may be interpreted as *the rule of the people, the voice of
the united people, the unity of the people*. This seems to be the Eastern concept of democracy. The Western concept seems to be more
active and challenging, while the Eastern concept is more passive
and conciliatory. These two concepts, different as they are, are cer-

tainly both legitimate.

The above theory makes it possible to predict some features of the future society as follows. As mentioned above, there will be two empires with the dual ideologies dominating the world and dividing it into two spheres of influence. The developed nations are supposed to elevate themselves to the status of demo-republics taking side with one of the empires with the appropriate extent of affinity. The developing countries will also strive to achieve the same status and take side with one of the empires. The two empires will struggle to expand their spheres of influence, managing at the same time the global issues, striving to soften their difference and adapt to each other, thus making the social (religious) life increasingly more private and the private life more social (religious). For that development to go smoothly, it is especially important to reform science and develop the universal religion and philosophy.

The conclusion about the tripartite structure of the Empire's government consisting of the Assembly, the Senate and the Church (or some other religious institution), may have important implications for modern society. One of the main inferences is that every of the above governmental institutions must also have a tripartite structure equally representing two other institutions. Such a tripartite structure seems to be the principal distinguishing feature of any self-governing society. That may suggest the way to solve currently frequent political crises by changing one-party and bipartisan governments to the tripartite self-governance.

Conclusion

There has been created the first systematic account of the science of politics that makes it possible to understand the true meaning of the known terms and facts, get rid of some delusions, predict in rough the future development of human society and choose the proper way out of its present crisis.

References

1. Lewis H. Morgan. *Ancient Society.* Archive of the Marxist organization, 1877.
2. Muhammad Hamidullah. *The First Written Constitution in the World: An Important Document of the Time of the Holy Prophet.* Lahore: Muhammad Ashraf, 1975.
3. R.K. Sinclair. *Democracy and Participation in Athens.* Cambridge University Press, 1988.

3 - LOGICAL FRAMEWORK OF ECONOMICS

Abstract

This part of the book proceeds with the work started by Karl Marx in developing economics as a logically consistent body of knowledge. There are three chapters: 1-Market production, 2-Capitalist production and 3-Monopolistic production. Chapter 1 shows how the individual labor develops into the market production with its Standard Market Supply (SMS) characterized by the Law of Value, the SMS being the essence of the whole theory. Chapter 2 shows how the market production develops into the increasingly sophisticated forms of capitalist production, from the handicrafts production to the industrial production and the industrial supply, the latter being the project of the SMS, an ideal form of market production. Chapter 3 shows the appearance of firms, self-financing production entities, and the centers, the initial forms of self-managing production. There are two kinds of centers with different orientations of interest: private-social (PS) and social-private (SP). The PS-center and the SP-center compete with each other and expand their markets by contracting unions of different liability with firms, making economy increasingly socialist. Finally there appear two global centers of PS- and SP-orientation of interest, competing with each other, streamlining their production, reducing prices and adapting to each other; they manage the global economy in the best possible way, presenting an increasingly perfect realization of the SMS project and transforming the socialist economy into the communal economy; the private interests becoming increasingly social and the social interests private. The evolution of economy is that of its three interdependent subsystems: production of goods, finance and trade. For a smooth transition to the socialist and communal economies, society needs ideological harmony.

Content

Introduction

Economics, in the proper sense of the word, is the science studying the social system of goods production. At present, however, economics is often considered an agglomeration of methods, models and theories dealing with various aspects of economy that can hardly be covered by a science deserving its name. This works is devoted to studying economics in the above strict sense of the word. Such a study is known to have been initiated in the 19th century by Karl Marx [1], but it was not and could not be completed, because of an underdeveloped state of science and the lack of systematic approach to science at that time, in particular. As we mentioned above, the systematic approach to science was developed first in philosophy by Hegel [2], and it was Karl Marx who showed how it worked in application to a concrete modern science. Now that such an approach has been understood, developed to a sufficient degree and successfully applied in exact sciences [3] and humanities [4], it is quite natural that we have endeavored to apply it to such a socially important science as economics. In the course of this work, there have been introduced new concepts and received new results. This work is a short account of the logical structure of economics and is expected to be discussed and enlarged by relevant examples and supplements.

Chapter 1. Market production

Economics is the branch of science studying the laws of production and exchange of goods in society. The element of any production process is *the individual labor*. Thus we should start the research with the study of the individual labor.

1.1. Individual labor

A. The individual labor is first of all a *specific labor* aimed at producing specific things for a specific consumption by human society. Therefore the individual labor is at the same time a social labor, *the* *labor in general.* Thus the nature of the individual labor is dual: it is a *specific-general* labor. The labor in itself is abstract, nonexistent; it becomes real only when applied to some material things, *the means of production.*

B. The means of production include *tools of production* and *raw materials* or *semi-products.* Similar to labor, the means of production are of a dual nature. On the one hand, they are the means supplied for a specific work; on the other hand, they are tools and materials in general. Thus the means of production are of *a general-specific* nature. In themselves, the means of production are abstract, non-existent: they become real only when acted upon by labor in *the process of production.*

C. The labor and means of production are realized in the process of production; the latter is the unity of those both sides in which they transit to each other, become inseparable, disappearing elements of that unity. The process of production is *real,* which manifests itself by *the real value* of *the goods* produced.

Comments:
There may be some objections as to the unreality of the means of production in themselves, because they usually appear to be some real useful things, for example, pens, computers, tractors for agriculture, etc. However, in putting such objection, one would overlook the fact that such things are not yet means of production; to become ones they must prove their ability for production in a real process of production, because otherwise they may prove unfit for it.

1.2. Production of goods

A. Taken directly, production is some technological process, *the production proper.* It is characterized by the average amount of labor and means of production spent for the production of goods, *the cost of production.* To re-stock the labor and means of production spent in the production of goods, the latter must be exchanged for other goods produced by other producers, which is the second side of the production, *the trade.*

B. During the trade, the product is assessed against the background of the whole production capacity of society; it is conditioned by the average conditions of society and thus acquires its social value, *the price.* The trade resumes the process of production, which thus becomes *a real production.*

C. The real production is mediated by society and becomes a social production, *a shop,* characterized by its *social status,* a composite value of its product, its *cost-price value.*

1.3. Production shop

A. One shop presupposes the existence of an indefinite multitude of shops. The specific character of every shop *separates* it from other shops.

B. Shops exchange their goods to resume production and therefore should *unite.*

C. These two contradictory trends of the shop, one for separation and another for unification, fall into unity giving birth to the society of private producers, *a simple market production.* The simple market production is an indefinite multitude of shops producing goods for exchange or sale in the market. As a unifying society, the market introduces a unified equivalent of goods – *money.*

1.4. Simple market production

A. The simple market society is a multitude of shops *identical* to each other because they all produce cost-price values. Therefore the simple market society is a *homogeneous society.* On the other hand, the market is a *discrete* multitude of isolated shops identified by their *trade marks.*

B. An isolated shop is also both discrete and homogeneous. It is discrete as one shop and homogeneous as identical to other shops of *the branch.*

C. Thus the shop proves to be similar to the market society because, in some respect, they are both homogeneous and discrete. With that similarity, the contradiction between the market and the isolated shop finds its solution in *the production branch* which is both a society of producers and a discrete producer of specific goods. As an individual producer, the branch has its representative, *a chief producer;* as a particular production society, it has *a code of rules, a charter.*

1.5. Branch of production

A. The branch is characterized first by the equivalent number of its producers, its *productivity.*

B. Any shop of the branch is to some extent different from its chief producer and may originate its own branch. Therefore any branch overlaps other branches and therefore has its *boundary* within itself, that is characterized by some kind of *elasticity.*

C. The branch with a limited productivity and a definite elasticity is a *definite branch.* It is characterized by a composite value, *productivity-elasticity.* The definite branch is a unified and united producer of specific goods and is able to establish *standards* of production.

1.6. Definite branch

A. The definite branch ensures regularity in production of its goods in constant conditions and, due to its elasticity, can adapt to changing trade conditions in the market.

B. In the market, the trade conditions are determined by the supply and demand relation which fluctuates about some average pattern and, therefore, is regular on average.

C. The market production, in which the regularity of the branch production is combined with the average regularity of trade, follows some stable pattern, *a standard market supply* (SMS), characterized by a relatively stable correlation of prices of different commodities.

Comments:
The stability of the market is the result of the play of its internal law, "an invisible hand", called *the law of value,* which regulates the correlation of different branches of production thus maintaining a relatively stable correlation of prices. The law of value adjusts the relation between the productivity of different branches through the spontaneous fluctuation of prices due to changes in demand and supply. Under the law of value, the goods are traded according to the cost of social labor necessary for their production. The SMS concept is *the essence* of the whole theory, the most fundamental concept of economics.

1.7. Summary

The logic of the theory stated in this chapter is outlined in Table 1, its cells containing the codes for the respective concepts according to the classification standard stated in Part 1.

Table 1. Market production

A Thesis	B Antithesis	C Synthesis	Q Quality
EC-1A1 Individual labor	EC-1B1 Means of production	EC-1C1 Production	EC-1Q1 Value of goods
EC-1A2 Production proper. Cost of production	EC-1B2 Trade. Price of goods	EC-1C2 Shop. Status of production	EC-1Q2 Cost-price value
EC-1A3 Multitude of producers. Isolation	EC-1B3 Exchange of goods. Cohesion	EC-1C3 Market produc- tion. Regulation	EC-1Q3 Unity of exchange. Money
EC-1A4 Society of pro- ducers. Identity of producers	EC-1B4 Individual pro- ducer. Identity of goods	EC-1C4 Branch of pro- duction. Chief producer	EC-1Q4 Charter. Authority
EC-1A5 Volume of branch. Productivity	EC-1B5 Inter-branch boundary. Elasticity	EC-1C5 Definite branch. Productivity- elasticity	EC-1Q5 Legal status
EC-1A6 Regularity and adaptability of definite branch	EC-1B6 Average regular- ity of supply- demand relation	EC-1C6 Average regular- ity of market. Standard Market Supply (SMS)	EC-1Q6 Law of value. Self-identity of market production

Chapter 2. Capitalist production

In Chapter 1 we investigated the general properties of market production and arrived at the conclusion of its self-identity expressed in
the formation of the Standard Market Supply. However that structure
of supply has not been expressed explicitly, and we should proceed

with its further investigation to achieve clarity in this respect.

2.1. Standard market supply

A. The standard market supply is regulated by the law of value according to which goods are produced and traded in proportion to the social labor necessary for their production. The law of value determines the proportion in goods exchange, sets definite relations between different branches, regulates the distribution of labor and means of production. Owing to that regulation, production in society is *self-consistent.* Thus the law of value is the law of market self-consistency.

B. The regulation function of the law of value operates via spontaneous fluctuations of prices due to changes in supply and demand. To be on the safe side, every producer works to accumulate certain amount of spare money. As a result, some producers manage to collect sufficiently large sum of money, invest it into production and get *profit.* In doing so, they turn money into *capital.* Capital is the foundation of *the capitalist way of production.*

C. The origin of capital enables its owner to acquire means of production, hire free workers and start production based on *hired labor.* The most primitive way of capitalist production is *a handicrafts production.* In such production the owner supplies the workers with tools and raw material and demands the production of goods for a definite payment.

2.2. Handicrafts production

A. The handicrafts production is the production of *identical* commodities made by individual isolated workers and collected by the owner who supplies them to market *in parties.*

B. The organized collection of market commodities produced by isolated workers suggests necessity for *an organized collection of* *workers* themselves.

C. The handicrafts production where workers are collected and organized in parties to perform the same production operations is *a cooperation,* a primitive organization of mass production.

2.3. Cooperation

A. The cooperation is based on a number of *identical* works. At the same time, any production has some *technological structure*, a *diversity* of different operations.

B. The diversity of operations suggests necessity for *the division of labor* to match the structure of the goods produced.

C. The cooperation modified by the division of labor turns into *a manufacture.*

2.4. Manufacture

A. The manufacture is based on a diversity of simple *mechanical* operations.

B. The use of simple mechanical operations suggests the possibility of using *machines* operated by *trained* workers.

C. The manufacture equipped with machines operated by trained workers is *a factory.*

2.5. Factory

A. The factory is characterized by the use of a number of machines, *a cooperation of machines,* with the workers *automatically* operating the machines.

B. The automatic character of labor at the factory presupposes the possibility of introducing *automatons* to perform automatic operations and the use of *qualified* workers to operate them.

C. The factory where the machines are substituted by a number of specialized automatons operated by qualified workers is *an industrial factory.*

2.6. Industrial factory

A. The industrial factory is the best solution for the technological side of production. The commodities produced must be *marketed.*

B. To market its product, the industrial factory organizes *an automatized collection* of its product, *delivering* it to the market, *selling* it to customers and *collecting the return.*

C. The industrial factory modified by an organized and automatized supply of its product turns into *an industrial supplier.*

2.7. Summary

Table 2. Capitalist production

A Thesis	B Antithesis	C Synthesis	Q Quality
EC-2A1 Self-consistency of market	EC-2B1 Sp Spontaneity of supply and demand	EC-2C1 Accumulation of capital. Handicrafts production	EC-2Q1 Foundation of the capitalist production
EC-2A2 Collection of identical products	EC-2B2 Collection of identical workers	EC-2C2 Cooperation	EC-2Q2 Primitive organization of production
EC-2A3 Diversity of operations	EC-2B3 Diversity of specialized workers	EC-2C3 Manufacture	EC-2Q3 Primitive systematic organization

EC-2A4 Mechanic labor	EC-2B4 Machines	EC-2C4 Factory	EC-2Q4 Primitive liberation of labor
EC-2A5 Automatic labor	EC-2B5 Automatons	EC-2C5 Industrial factory	EC-2Q5 Advanced production. Labor liberation
EC-2A6 Advanced organization of production	EC-2B6 Advanced organization of trade	EC-2C6 Industrial supplier	EC-2Q6 Project of SMS realization

Chapter 3. Monopolistic production

In Chapter 2 we investigated the development of the capitalist way of production and arrived at its highest form – the industrial supplier. The latter produces goods and supplies them to market. The industrial supplier has the best organization of both stages, producing and marketing the goods, and the best *ideal* realization of the Standard Market Supply. However, the structure of the industrial supplier is not articulated, not developed to the level of system, because it is focused on producing one or a restricted number of commodities. For that reason, the industrial supplier is not a practical solution of the SMS, but rather its *project*. So we should proceed with the further investigation to see how that SMS project is implemented.

3.1. Industrial factory and supplier

The industrial factory and industrial supplier both *alternatively* produce and market goods. The difference is that, while the industrial factory has the marketing stage incorporated *implicitly*, in the industrial supplier it is incorporated *explicitly*. These two forms of production present dual images of one another and can turn into one another, which suggests the existence of their unity, a *self-financing* entity, *a firm*. In the firm, in contrast to its constituents, the processes of production and marketing are uninterrupted. The firm has its distinguishing trade mark, an advanced position among competitors and a certain *market power.*

3.2. Firm

The firm is able to produce goods and supply them to market. Each of its two major parts, industrial and commercial, performs two functions of its own: the industrial part is responsible both for the production of goods and the modernization of equipment, while the commercial part is responsible both for marketing the goods and financing their production. To ensure an efficient and smooth functioning, all these parts need permanent attention which the firm is unable to provide. Indeed, when, for example, it is engaged in modernizing equipment, it should stop or cut the production. Thus the firm cannot work efficiently on its own and seeks to double its structure or merge with another firm; when achieving that, the firm turns into *a center.* The latter ensures a permanent production of goods and their supply to market; it becomes *a leading producer* of the branch, acquiring a significant market power and *a local governing authority.* The center needs a distribution of management involving some employees and, therefore, presents an initial form of *self-management*, the principal criterion of the *socialist* way of production.

3.3. Center

The center is a *commercial* enterprise supplying its products to customers, which is its *social* function; so the whole function of the center is dual, *private-social.* That duality suggests necessity for enterprises with a dual function, a *social-private* one, and therefore two kinds of economy in general: one with the private-social (PS) orientation of interest, another with the social-private (SP) one. Their difference is that the PS-economy is more active and extrovert, while the SP-economy is more passive and introvert. So every branch of economy may ideally contain, at least, two leading centers with the dual orientations of interest, a PS-center and a SP-center. The rise of dual monopolistic centers marks the beginning of the dual monopolistic – *socialist* - economy. Thus, achieving its extreme development in the industrial supply, the capitalist economy transforms to its opposite – a socialist economy.

3.4. Evolution of the center

The centers grow by making associations with firms of different orientations of interest. There are *solidarity* firms that *inherently* have certain orientations of interest, *preference* firms that *prefer* certain orientations, and *neutrality* firms that have no certain orientation of interest. So, there are five kinds of firms to make agreements with, and therefore six kinds of association for each center.
The evolution of the center is actually the evolution of its three interdependent subsystems responsible for the production of goods (the Specific), for its financing (the General) and the trade (the Unity), every two of which suggesting the third.

The center of the first level, the 1-center, makes solidarity associations and preference associations with the firms of the respective orientation of interest, as well as neutrality associations, thus rising successively to higher levels and becoming the 2-center, the 3-center and the 4-center, respectively, absorbing, possibly, on that way other centers of the respective interest orientation. Then, in the same manner, the 4-center contracts whatever possible agreements with the preference and solidarity firms of the opposite orientation of interest, thus rising to the 5-center and 6-center, respectively.

Finally, there emerge two 6-centers of opposite orientation of interest; they make agreements with each other, thus rising to the global centers, the 7-centers. Now the world economy has achieved its highest possible level of concentration and monopolization and is divided into two economies of different interest orientation, the private-social, PS-economy, and the social-private, SP-economy, competing with each other.

To withstand competition, the global centers streamline their organization, keeping the prices at the lowest possible level and increasingly involving the working personal into their management, thus gradually making the whole economy *self-managing*. The global centers acquire *political* power and become global political centers as well.

3.5. Self-managing economy

With the formation of the global centers, the socialist economy achieves its highest possible development becoming self-managing: the global centers becoming the societies of producers that work for the whole community. At this stage, the dual purpose of economy is fully realized, with the private interests becoming social and the social interests private. So this stage of economy presents the best possible realization of the SMS project. The socialist economy now turns into *a communal economy* where production is performed not for the sake of profit but for the sake of production itself, for the sake of useful labor which has become free, containing the law of value as *a moral necessity*.

3.6. Summary

The logic of this chapter is outlined in Table 3, where the first column shows the evolution of the PS-center, from the PS1-level to the PS7-level; the firms involved in associations are indicated in columns A-G (A- PS-center proper, B – Solidarity PS-firms, C – Preference PS-firms, D – Neutrality PS-firms, E – Preference SP-firms, F – Solidarity SP-firms, G – SP-center proper), while the quality of the respective centers is shown in column Q; the cells of the table contain the codes for the respective concepts. The quality of the center corresponds to the quality of its association, except for the 5-center and 6-center whose qualities are called tentatively 'Expansion-5' and 'Expansion-6', respectively.

3.7. Discussion

The above theory suggests the way of solving economic crises and the management of the global economy at its highest stage. The scheme of the global management of economy at this stage, following the bottom row of Table 3, is shown in Fig.1. The centers PS7 and SP7 compete for global markets thus adapting to each other and thereby managing the global economy in the best possible way. When one of the center becomes dominant and increases its market power by supplying goods, for example, at lower prices than its counterpart, some of its associated firms, seeing injustice and worsening conditions of their staff, may change sides making association with the counterpart center thus restoring the balance of global power. That example suggests, in particular, that maintaining the dual character of economy is the best possible way of solving global economic crises.

Table 3. Monopolistic production

Firms → Centers↓	A PS1	B Sol	C Pref	D Neut	E Pref	F Sol	G SP1	Q Quality
EC-3-1 PS1-center	EC-3A1							EC-3Q1 Dual economy
EC3-2 PS2-center	EC-3A2	EC-3B2						EC-3Q2 Solidarity
EC-3-3 PS3-center	EC-3A3	EC-3B3	EC-3C3					EC-3Q3 Preference
EC-3-4 PS4-center	EC-3A4	EC-3B4	EC-3C4	EC-3D4				EC-3Q4 Neutrality
EC-3-5 PS5-center	EC-3A5	EC-3B5	EC-3C5	EC-3D5	EC-3E5			EC-3Q5 Expansion-5
EC-3-6 PS6-center	EC-3A6	EC-3B6	EC-3C6	EC-3D6	EC-3E6	EC-3F6		EC-3Q6 Expansion-6
EC-3-7 PS7-center	EC-3A7	EC-3B7	EC-3C7	EC-3D7	EC-3E7	EC-3F7	EC-3G7	EC-3Q7 Global center

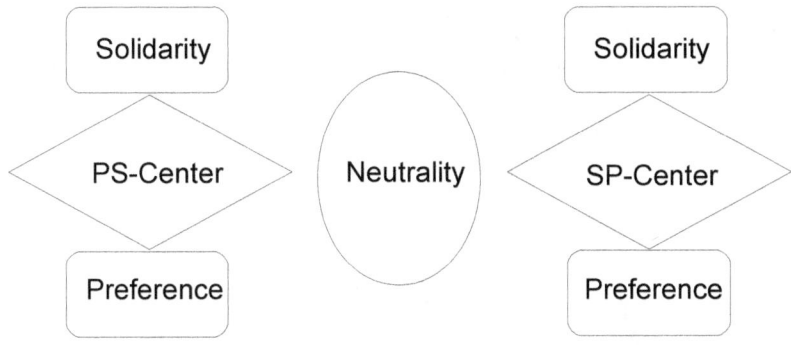

Fig. 1. Global management of economy

Conclusion

The above work completes in rough the development of the logical structure of economics started in the 19th century by Karl Marx. It shows the way the world economy will most likely follow and suggests likely measures to remove possible obstacles in that way.
The main result is that the monopolistic economy transforms naturally, without any revolutions, through the growth and competition of dual centers, first into the socialist economy and then into its highest stage - the communal economy.

References

1. Karl Marx. Capital, vol 1: A Critique of Political Economy. Amazon.com.
2. G. Hegel. The Logic. Encyclopedia of the Philosophical Sciences, Vol.1. Clarendon Press, 1874.
3. Igor S. Makarov. A Theory of Ether, Particles and Atoms. Second Edition. 2010. ISBN-13: 9 781441 478412 (www.amazon.com). Online: http://kvisit.com/S2uuZAQ.

4. Igor S. Makarov. The science of politics.
Online: http://kvisit.com/ShdDFAQ.